CRTD-VOL. 28

RESEARCH NEEDS AND OPPORTUNITIES IN FRICTION

Report of
The Workshop on Friction

Sponsored by
The National Science Foundation

Transcribed from
The Reports of the Session Chairmen by
Traugott E. Fischer

THE AMERICAN SOCIETY OF MECHANICAL ENGINEERS
UNITED ENGINEERING CENTER / 345 EAST 47TH STREET / NEW YORK, NEW YORK 10017

This report is based on a workshop funded by NSF Grant MSS-901660
to the ASME Board on Research and Technology Development.
NSF Program Director, Lallit Anand

ISBN No. 0-7918-1202-2

Library of Congress Catalog Number 94-71043

CONTENTS

SUMMARY

This report presents research recommendations in the field of friction. The recommendations are based on the analysis of three types of information:

- a survey of the industries in which friction plays an important role
- selected industrial developments that depend on progress in the control of friction
- a review of recent basic scientific developments in the field

Findings indicate that friction impacts a large number of applications and that progress in its control will provide energy savings as well as producing other effects:

- enabling important design advances
- increasing product quality and production efficiency
- contributing to health and safety and the performance of recreational equipment

As common in scientific pursuit, a partial match between technical needs and scientific advances has been found. However, certain industrial problems remain without a recognizable solution and some scientific advances do not yet have foreseeable applications.

Application-based research needs and the present research opportunities are as follows.

Research Needs

- extension of the temperature range for friction control
- friction control (lubrication) in difficult chemical environments
- environment friendly lubricant
- lubrication adapted to novel materials
- control of friction in high-density magnetic recording and micromachines
- better knowledge of the role of surface topography in mixed lubrication
- low-friction unlubricated surfaces
- mathematical models of friction usable in machine design
- relationship between friction and machine dynamics
- more effective transfer of tribological principles to the engineering community

Research Opportunities

- fundamental science of energy dissipation processes and interatomic force potetials
- molecular dynamic simulations of friction, wear, and adhesion
- atomic and molecular friction and lubrication experiments
- adaptation of new surface analytical tools to tribology tribochemistry
- prediction and generation of tribological films
- design and application of new tribomaterials
- engineered interfaces
- relationship between macroscopic and microscopic viewpoints
- surface dynamics (behavior of asperities in sliding, interaction of vibrations and friction)

It is the recommendation of this report that this interdisciplinary interaction in the study of friction be nurtured, and viewed as a scientific priority.

PREAMBLE

This report analyzes current needs and opportunities for research in friction. The methodology employed is as follows.

(1) The industrial problems in which friction plays an important role were studied and attempts were made to identify the needs for progress in the understanding and control of friction.

(2) The present status of our understanding of friction was assessed, devoting special attention to recent advances and the areas in which progress is most probable at this time.

(3) A match was attempted between the technological pull and the scientific push.

A number of these matches have been identified, but obviously, the nature of progress is such that a perfect match between needs and innovation is not possible. Certain technological problems have remained unsolved for a long time and we still do not know how to solve them; and certain advances have been achieved in the fundamental science for friction, for which a practical application is not apparent at this time.

A prudent research program thus will have three main components:

- applied research aimed at solving problems with present science
- the search for methods to solve the important technological problems
- the nurturing of basic scientific progress

The information and recommendations presented here are the results of a workshop held in Washington, D.C., June 24–26, 1992, to which approximately 40 experts were invited, representing as full a range of orientations and interests as could be assembled. The program and list of participants are provided in the appendix at the back of the book. In preparation for the workshop, Marshall Peterson, Lisa Tomanio, and Howard Clark collected information on the industrial activities in which friction problems play a significant role. The workshop was divided into four sessions. In the first, David Tabor gave an overview of the concepts and phenomena of friction, L. Irwin Singer presented a synopsis of the NATO Advanced Summer Institute's study on friction (held in Braunlage, Germany, in August 1991), and Marshall Peterson presented the findings of his survey. The second session was split into five groups that explored the technological needs for progress in the areas of mechanical systems, materials processing and manufacturing, data processing and micromechanics, road and rail friction, and general applications. Each group identified the progress needed from the viewpoint of the application and translated these needs into tribological research areas. The same organization was used for the third session, in which five groups examined the status of knowledge and the opportunities for further progress in fundamental theories, experimentation at the atomic and molecular scale, materials phenomena from a microscopic viewpoint, friction phenomena from a continuum viewpoint, and surface chemical phenomena attending friction. The final session endeavored, first in three groups and finally in full session, to achieve a unified picture and formulate broad recommendations for future research in the field.

Friction, as opposed to wear, was selected because it has been less thoroughly investigated recently. In order to provide a focus on the basic phenomena, the design of hydrodynamically and elastohydrodynamically lubricated bearings and the chemical formulation of lubricants were not considered in this study. However, the mechanisms of energy dissipation in very thin liquid layers, where novel insights have been gained recently, and the action and design of boundary lubricants are included.

In order to present a coherent picture, we will organize the material according to the concepts rather than to their chronological presentation. The talks of Drs. Tabor and Singer are not reproduced here since they are available in the book based on the study at the NATO summer institute.[1] However, their conclusions and recommendations are included in the main text of this report.

---

[1] L. I. Singer and H. M. Pollock, "Fundamentals of Friction: Macroscopic and Microscopic Processes," NATO ASI Series E: Applied Sciences, Vol. 220, Kluwer Academic Publishers, Dordrecht, Boston, London (1992).

Application Areas of Friction

The information of friction problems in industry, assembled by Peterson, Tomanio, and Clark, was gathered from a survey of the technical literature, direct information provided by members of the American Society of Mechanical Engineers (ASME), and from responses, by industrial engineers, to published requests for information

Progress in friction control will have diverse benefits:

- a reduction to the lowest possible friction, for smooth operation or energy conservation
- an extension of the temperature range in which effective lubrication can be provided
- the achievement of maximum traction
- a better knowledge or prediction of friction values
- the reduction or utilization of friction vibration
- the elimination of stick-slip or static friction after long storage
- lubrication schemes for novel materials and novel lubrication schemes for extreme environments

Friction plays a role in many industries. Its better control will help us (1) conserve and exploit our energy resources, (2) increase our competitiveness by improving product quality or manufacturing efficiency or by stimulating novel technologies, and (3) improve personal safety and health.

Energy savings will be achieved by better control of friction through the direct decrease of frictional losses, but more importantly, by enabling changes in the design and operation of machinery, for instance in the following applications:

- engines and gas turbines, rotating or sliding seals (higher efficiencies and lower weights can be achieved with higher speeds and higher operating temperatures)
- power transmission (through lower energy losses in belts and gears, and by the construction of continuously variable transmissions that allow power plants to operate in their most efficient regime)
- traction in railroads (through decreased rolling friction)
- oblique drilling in prospecting and production of gas and oil (increased drilling distances can be achieved through lower friction)

Increased competitiveness will result from better product quality and operation in the following technologies:

- control systems and robots (smooth operation and the avoidance of stick-slip at low velocities is important for operation quality)

- space equipment (operates in vacuum and must be able to move smoothly, often after prolonged storage or immobility)
- textile and fiber manufacturing (achievement of controlled yarn tension is essential)
- brakes, clutches, friction materials (control of friction is required for smooth operation)
- metal working control of friction is necessary for the efficiency of production and product quality)
- material cutting and machining
- friction welding
- polymer processing
- powder processing
- sand and rock handling
- foundations (construction and stability of piles depend on friction against the soil)
- textile and fiber applications (feel and comfort of the textile is influenced by friction)
- sports equipment (performance, quality)

Novel technologies or step-out improvements will be made possible in the following technologies:

- magnetic recording progress in recording density requires very small gaps between recording medium and apparatus, thus preventing the hydrodynamic lubrication by air)
- novel materials (requires the development of material specific methods of friction and wear control)
- micromechanical systems (an emerging technology for which no efficient friction control exists at this time)

Friction plays an important role in health and personal safety in the following applications:

- medical prostheses
- shoe/floor traction
- control of frictional vibrations and noise (noise pollution)
- road and tire traction
- tire traction on ice and snow
- earthquake control

Research Needs in Technological Applications

In this section, several industries are analyzed in which progress in product performance and quality, production efficiency, energy conservation, environmental compatibility, safety, or other considerations depend critically on a better control of friction. The technological demands or advances are analyzed in an endeavor to identify the friction research needed to achieve the required advances.

● ● ● ● ● ●

MECHANICAL SYSTEMS

Robert A. Harmon, *Chairman*

Mechanical systems — engines, machines, automobiles, airplanes, and other transportation equipment — are an obvious target for friction research. Routine friction problems are well handled by existing lubrication technology and further energy savings in these cases will be marginal. Experience has shown, for instance in the case of the automobile, that larger energy savings can be achieved by changes in the design and operating parameters of the equipment than by the decrease of friction alone, and that these changes often depend on novel methods for friction control. The design trend of almost all machines is toward smaller size and higher operating speeds. In power plants, such as internal combustion engines or turbines, the emphasis is focused on the increase in operating temperatures to achieve higher thermodynamic efficiencies.

In the case of piston and gas turbine engines, the principal issues driving present technical development are fuel savings (increased efficiency and the utilization of alternate fuels, such as gas and alcohols), emission control (by higher efficiencies and changes in lubricant chemistry), design changes, and the utilization of novel materials, such as polymers, composites, and ceramics to decrease cost and wear or reduce total weight.

For robots, the principal issue is smooth operation (i.e., the absence of stick-slip in slow motion). Technical challenges are presented also by systems in which the application of conventional lubrication is restricted (such as in food processing) and by machinery working in extreme environments, such as in space or at very low temperatures.

These technologies are dependent on the progress in the science of friction in several directions.

A major development in the area of diesel engines is the increase in operating temperatures in order to achieve higher Carnot efficiency. This points to top ring reversal temperatures in the order of 1000°F to 1200°F. Successful uncooled operation of ceramic turbines at 2500°F turbine inlet temperature means higher regenerator inlet temperatures. Current technology is only good to about 1800°F to 1850°F. This tribology problem (high-temperature lubrication) is pacing the development of advanced diesel engines and placing undue constraints on passenger car engines. Since hydrodynamic lubrication by liquids or gases remains the most effective means for reducing friction and avoiding wear, this amounts to defeating the temperature-dependence of the viscosity of fluids and controlling the adsorption properties and self-organization of boundary lubricants over a large temperature range.

The problem is twofold: to provide a liquid lubricant with sufficient thermal stability and viscosity, and to provide a boundary lubricant (friction modifier) that remains stable on the surface at these temperatures. Present friction modifiers operate at temperatures below 120°C. Synthetic hydrocarbon lubricants have been developed for gas turbine engines, but the temperatures they are capable of limit the development of the engines. Novel gas-delivered lubricants (Klaus, Lauer) are being explored but need further development. These provide boundary lubricants only with a short lifetime, but are replaced continuously by a gas carrier. These schemes are still in the experimental stage, and they cannot provide the hydrodynamic lubrication necessary for low friction and wear.

In high vacuum, in outer space, at very low temperatures, in certain food or pharmaceutical processes, machines must function reliably without the benefit of conventional lubrication. Solid lubricants are employed to solve many of these problems, but they are suitable only for a limited amount of sliding. The avoidance of stick-slip friction is important in some of these applications; in this case, a better understanding of mixed lubrication and the development of low-friction surfaces is required.

Alternate fuels, such as methanol, compressed natural gas, and hydrogen, are being explored in order to utilize the natural resources more fully and reduce pollution. They lack the lubricating properties of liquid fuels and can affect the life and reliability of fuel system components, injection pumps, and valve train components. This will require the use of different materials, the addition of nonpolluting lubricants to the fuel, or other schemes.

Environmental demands and regulations require nonpolluting (i.e., ashless) lubricant additives. The scientific challenge consists in the development lubricant molecules that do not contain elements forming pollutants in combustion, and that remain stable and adhere to solid surfaces at higher temperatures than present technology allows.

The use of novel materials (ceramics, composites, polymers) require special consideration for the development of boundary lubricants and other additives. Their successful application requires a better knowledge of their tribological behavior in order to establish procedures that are adapted to their unique physical and chemical properties; the direct transfer of technology that is appropriate for metals often leads to disappointing performance.

Summary of Research Recommendations

High-Temperature Lubrication

- development of high-temperature lubricants with suitable chemical stability and very high-viscosity index (i.e., low-temperature dependence of viscosity)
- development of novel boundary lubricants that remain adsorbed and active to higher temperatures than present technology (i.e., higher than 110°C)
- further development of novel lubricating schemes with gas transport of boundary lubricants or formation of lubricants by chemical reaction of gases with surfaces
- development of novel lubrication schemes
- supporting science: surface chemistry, tribochemistry, liquid rheology
- identification and assessment of various avenues to high-temperature operation without sacrifice of safety or reliability

No Stick-Slip Operation in Space, Robots

- development of novel solid lubricants with longer life
- development of low-friction surfaces
- development of novel boundary lubricants
- supporting science: basics of friction, surface science

Alternate Fuels, Novel Materials

- development of nonpolluting lubricant additives to fuel
- tribochemical reactions of lubricants with solid surfaces
- supportive science: surface chemistry, tribochemistry

Nonpolluting Lubricants

- design of lubricant additives that do not form pollutants
- high-temperature chemistry
- supportive science: chemistry

● ● ● ● ● ●

MATERIAL PROCESSING AND MANUFACTURING

Harmon Nine, *Chairman*

Advances in material processing are essential to the competitiveness of our industry, as they impact the economics of manufacturing and product quality. In addition, the utilization of the novel materials that have been developed in the recent past depends critically on the development of processing techniques that will allow their production at acceptable prices. Each material processing technology has its own challenges and will benefit from progress in friction control in its own way. Metal rolling depends on friction for its operation, and the required surface properties are obtained by the use of lubricants; in this situation, the roughness of the surfaces influences friction in a manner that is not sufficiently controlled. In the processing of polymers and composites, friction is involved in heating the resin, in the flow of the material in the mold, especially when fiber-reinforced composites are processed, and in the release from the mold. The processing of high-quality ceramics depends on the flow and packing of the powders; the latter is influenced by friction between particles; the same holds true in powder metallurgy. In all machining, friction, stick-slip, and vibration influence the chatter of the tool and, therefore, the precision of the workpiece and the quality of its surface. Finally, realistic models of friction must be developed for computer modeling of material processing technologies.

Surface roughness is an important consideration in metal forming on two counts: it is often a specification in its own right, and it influences friction between tool and workpiece. In a number of processes, friction is a critical factor in the ability to form a part. In metal rolling for example, friction between the rolls and the workpiece is essential. Present research needs are focused on the development of controlled surface roughness (texturing) of the tool in order to achieve the desired friction forces in the presence of the lubricant. In extrusion, friction of the material against the dye determines the flow of the material, the energy needed for the process, and the materials properties of the extruded part. Since conventional profilometer measurements, such as the Ra and peak count, have not shown the capability of predicting friction behavior, the characterization of surface roughness in a way that shows the expected correlation to friction needs to be found. New methods, such as fractal geometric approaches, must be pursued to ensure that roughness can be controlled to give optimum frictional characteristics.

In polymer processing, pellet-fill dynamics is controlled by friction; thus an understanding of friction is important to the quality and economics of the manufacture of polymer parts. Friction will also impact polymer manufacturing in affecting the release of the parts from the mold.

The control of the flow and packing of powders, in powder metallurgy and the processing of ceramics, shows great potential for improvement. Increased productivity, reduced cost, and improved material quality will be the benefits. The role of interparticle friction in powder flow for optimum compaction requires investigation of the relevant tribological factors.

The high cost of machining and its limited reliability, due to surface defects, still present a formidable obstacle to the widespread utilization of ceramics, which otherwise offer the obvious benefits of chemical resistance, high-temperature stability, and low density. An understanding of friction at the tool/workpiece is required to control energy consumption, optimize surface quality, and increase tool life in the machining of ceramics and other advanced materials. While energy consumption and tool life influence the cost, surface quality has a large influence on component reliability.

A great deal of activity in computer modeling of metal forming is underway in private industrial, academic, and government groups around the world. The potential cost savings obtained by computer modeling, as opposed to trial-and-error methods, is enormous. Proper representation of friction in these models is vital to their success. Accurate friction models need to be developed and codes may have to be modified to best exploit these models.

Machine vibration and chatter decrease precision and surface quality. Tribosystems consist of dynamic assemblies that interact through an interface comprised of asperities and debris. The whole system vibrates in various modes, observed as macroscopic machine chatter and noise. The vibrations are important globally in influencing variations in the normal and frictional forces; they are significant locally at the asperity level in influencing interactions and contact mechanics at the interface. Thus, generic tribodynamic modeling is necessary for a complete description of local contact conditions and resulting frictional behavior.

Many current lubricants used for material forming and machining have polluting components. Components such as chlorine must be replaced to make these lubricants environment friendly. Air pollution from lubricants evaporated and burned in hot forming is an important problem. If research does not make it possible to modify the lubricants, significant costs will be incurred for hazardous waste disposal methods.

Summary of Research Recommendations

Surface Roughness

- novel, more accurate mathematical characterizations of surface topography
- elastohydrodynamic lubrication as a function of surface topography

Polymer Processing

- polymer-metal friction
- low-adhesion surfactants
- supporting science: surface adhesion

Powder and Ceramic Processing

- hydrodynamics of powders
- development of boundary lubricants for powders
- supporting science: surface chemistry, mathematical modeling

Ceramic Machining

- more detailed investigation of creation of subsurface defects as a function of depth of cut and cutting speed
- chemomechanics of cutting, chemically assisted machining
- development of alternate (tribochemical or other) machining methods
- supporting science: tribochemistry, micromechanics, wear

Computer Modeling

- more accurate determination of the dynamics of friction, origin of vibrations, effect of vibrations on friction
- basic friction theory

Machine Chatter

- study of the interaction of friction dynamics and machine dynamics
- friction dynamics: origin of vibrations, effect of vibrations on friction

Nonpolluting Lubricants

- See *Mechanical Systems*.

● ● ● ● ● ●

DATA PROCESSING AND MICROMACHINES

Clarence J. Spector, *Chairman*

The storage and retrieval of data depend on the relative movement of read-write head and storage medium at very small distances and small forces between the two. The most important tribological problems in this industry are those that might be second order in many other industries. The present technology relies on hydrodynamic lubrication by an air film separating the head and the medium, and utilizes deposited boundary lubricant for protection in start and stop operation. Projected developments towards higher data storage density depend in a large measure on a decrease of head-to-medium distance, which is not compatible with hydrodynamic lubrication and requires novel, yet unidentified technologies for the control of friction. The following are some specific technological needs: requirements for extremely low-power start and run for storage systems; microinch precision, and low overshoot in position accessing of data tracks.

0 to 5 μ in. separation between the surfaces in relative motion and relative velocities ranging from micrometer to tens of meters/second. Decrease of this separation to the smallest possible values is an important condition for the achievement of the high-storage capacities that are theoretically possible.

There are strong distinctions between static and dynamic friction requirements. Static friction takes the form of stiction in these machines. Stiction is caused by the surface tension of a meniscus formed by applied or condensed liquids. Stiction is important in systems with very low forces between the surfaces; it has important time dependency. Microinch surface separation provides the basis for meniscus force dominance at rates dependent on surface diffusion of the lubricant.

Micromachines constructed with the methods of silicon photolithography present interesting possibilities that are currently being explored. The application of liquid or classical solid lubricants to these systems is not possible because of their microscopic scale. Although the problem can be alleviated by judicious design where relative movement is accommodated by rolling, the challenge of their effective lubrication is still not met. The development of novel methods for the evaluation of interface interactions is required for micromachine scale devices.

All these problems require friction control with nontraditional means, without the use of liquid or gaseous layers separating the surfaces. Because of the small dimensions and the small forces involved, adhesion and condensation effects become important. There is a need for research in boundary layer formation, surface characterization, and design of molecules with appropriate end groups, controlled entanglement, and self-healing after being damaged by physical, electrical or chemical stress. Programs directed at the design of interfaces should be extended by coupling to an organic molecules synthesis community as is done in Europe and Japan.

It is doubtful that boundary lubrication by adsorbed molecules is the solution to the problem of ultrahigh-density storage because the friction coefficients ($\mu = 0.05$–0.1) are still too large for high-speed operation, in which very low power consumption is essential and stiction forces will be prohibitive for successful start and stop operations. A new technology, based on entirely novel concepts, will need to be developed.

Summary of Research Recommendations

- search for a novel lubrication scheme that allows very close contact or low-friction surfaces (presently an ARPA research initiative)
- tribochemical formation of boundary layers
- design and synthesis of novel, predeposited boundary lubricants
- supporting science: basic theory of friction, surface chemistry, tribochemistry

● ● ● ● ● ●

ROLLING CONTACTS – ROLLING BEARINGS, ROAD, RAIL, CONTINUOUSLY VARIABLE TRANSMISSIONS

Marshall Peterson, *Chairman*

There is continued industry interest in gaining a better understanding of the traction in rolling contacts. Usually, the aim is to obtain maximum traction in the rolling contact, coupled with the smallest possible energy dissipation and wear. The two requirements are mutually contradicting because they both involve adhesion between the two surfaces. In any rolling contact, the interface consists of both slip and non-slip regions. Significant advances have been made in predicting and measuring the respective locked and slip regions. The amount of microslip can be accurately measured and used to estimate wear and motion of the contacting bodies; however, very little is known of the microslip itself. We do not know, for instance, whether the same principles of friction, wear, and lubrication apply when sliding distances are less than 1 m in a rolling contact, nor what the range of elastic and plastic response is, or whether microslip is a static or dynamic phenomenon. To answer these questions a broad program is recommended with the following elements.

Summary of Research Recommendations

- A suitable friction device should be constructed that allows motions in the 0.01 m–1 m range at a variety of loads. Friction should be accurately measured.
- Friction mechanisms should be studied as a function of rolling distance. Friction and wear values should be compared with values obtained in long distances.
- The transition between static and dynamic friction should be investigated with the aim of determining when the transition occurs and how much elastic and plastic deformation of the surfaces is involved.
- Lubricated friction should be investigated. It would be useful to know whether the amount of metal contact through the lubricant film is a function of the length of microslip or only a function of load and rolling velocity.
- Analytical investigations should be conducted to determine if run-in occurs during microslip, namely whether there is sufficient deformation to allow run-in.
- Finally, friction control techniques should be investigated. The best method of lowering friction and preventing surface damage under microslip conditions while maintaining traction should be developed.

● ● ● ● ● ●

GENERAL APPLICATIONS

Frederick F. Ling, *Chairman*

There are many technologies in which progress in the control of friction will have various benefits, such as an increase in safety, comfort, manufacturing efficiency, product quality. Some examples are discussed below.

Garment Construction. Friction considerations appear in many ways in the vast arena of garment construction. One example is the judicious use of friction knowledge for handling fabric during the processes of joining by sewing machine. Friction is also important in the feel of garments.

Prosthetic Devices. Friction problems associated with devices ranging from hip joints to heart valves determine the satisfactory functioning and the life of the device. In addition, nature employs lubrication techniques adapted to the reciprocating movement of joints: a better understanding of these processes could lead to their adaptation to technical problems.

Sliding Electric Contacts. This is an important area both on the earth's surface and in space. There is the need to transmit low-noise signals, and low as well as high currents across slip rings with low friction, often in environmentally adverse conditions.

Composite Materials. The mechanical properties of fiber reinforced composites depend on the friction between the fibers and the matrix. When the role of the fiber is to strengthen the material, as in polymer matrix composites, the friction between matrix and fiber must be high. When increased toughness is sought, as in ceramic matrix composites, the friction must be moderate and controlled: enough sliding must occur near the propagating crack tip to prevent fracturing the fiber, but friction must be large enough to hold the matrix and prevent it from fracturing.

Sports and Recreational Equipment. A large number of friction related problems are present in this area, from friction reduction and increased performance to the achievement of a specified friction coefficient for traction and safety.

Tire and Shoe Traction. While the science and technology of tire friction on roads are well developed, further progress would be possible by the application of novel concepts. The development of tires and comfortable shoe soles with increased traction on ice is a challenge that has not been met to date.

Environmental Protection. The development of nonpolluting lubricants is being mandated by law. While this is a chemical problem, it impacts friction research because of the need for novel boundary lubricants.

Sensors. The development of sensors for frictional contact stresses and local temperatures would contribute immensely to the control of machinery, the development of lubrication schemes, and basic research in friction.

Summary of Research Recommendations

- friction between fibers and metal, fibers and fibers
- study of natural lubrication in animals
- boundary lubricants compatible with the human body
- study of tribochemial lubrication possibilities in vivo
- friction and friction enhancement on ice
- development of friction sensors
- supporting science: fundamentals of friction, tribochemistry, adsorption, biology

Scientific Progress and Opportunities

In this second section of the review, the current state of knowledge in friction, the recent progress achieved in the science and the most likely directions in which further advances can be expected will be examined. Recent developments in scientific thought and techniques have had their impact on tribology. The advent of high-power computing has permitted detailed view into contact and friction phenomena through computer experiments (that allow us to see the phenomena on an atomic scale without necessarily providing insight into causal relationships); novel scanning probes (scanning tunneling and atomic force microscopes) have permitted experimentation with artificial single asperities and imaging of surfaces with atomic resolution; solid state theory has progressed to the point where the difficult problem of energy dissipation in friction, from the most fundamental viewpoint, is being revisited; mathematical approximations (in particular the finite element method) permit the analysis of macroscopic problems with sufficient complexity to become realistic; and surface science has developed a number of optical methods that open the possibility of observing adsorption and chemical reactions of surfaces of solids covered with a liquid (where the powerful electron and ion spectroscopies could not be applied). Some of these new developments have changed our views of friction, others have confirmed what is already known. Many are in an early state of development and need further work – highly recommended for the promise of real progress. The advances achieved and the most promising avenues for further progress, as they appeared to the participants in the workshop, are organized by the scale of the phenomena, namely the atomic and molecular, the microscopic and the macroscopic viewpoints. In addition, the fundamental theories and the surface chemical work have received special sections. A major challenge, in the research of frictional phenomena and applications, is to build bridges between these different activities, which are often housed in different scientific communities, and achieve a unified picture of the phenomena and the problems.

● ● ● ● ● ●

FUNDAMENTAL THEORIES

Mark Robbins, *Chairman*

The basic goal is to develop an understanding of how, why, and where energy dissipation occurs when two solids slide past each other. This understanding is expected to reveal which properties of a solid and which types of couplings between the solids lead to high or low friction. Different energy dissipation mechanisms come into play at different scales. The basic mechanisms, the scientific questions, and the theoretical tools will differ when one considers friction from the viewpoint of the macroscopic, the intermediate, or the atomic scale.

At the smallest scale, fundamental treatments of electron and phonon excitations and interatomic forces can examine the origin and size of friction forces as a function of the basic material properties.

At the intermediate scale, molecular dynamics calculations simulate atomic motion using approximate interatomic interaction potentials. Such calculations are capable of simulating adhesion and microscopic wear processes, effects of high temperature, and the behavior of liquids confined in very thin layers between the surfaces.

At the macroscopic scale, the continuum viewpoint is most appropriate. The relevant questions are whether the relative motion is accommodated at the interface of the two bodies in relative motion, inside these bodies (for instance by plowing or plastic shear) or in a third body (a solid, liquid, or gaseous lubricant or wear debris) placed between the two. These theories model friction by using empirical descriptions of elasticity, plasticity, viscoelasticity, hydrodynamics, and energy dissipation. The friction between the surfaces is generally described by empirical or arbitrary parameters. The most powerful mathematical tool at the moment is the finite element approximation.

While different individuals may be involved in studies at each of these three scales, it is apparent that there is a great need for cross fertilization. For example, quantum mechanical calculations may provide the potentials needed for molecular dynamics simulations which may, in turn, provide the constitutive relationships needed for continuum theories. The concepts brought out at larger scales will shape the questions asked at smaller scales. As always, experiments will play a key role in guiding assumptions and testing conclusions. One may also learn much from work by surface scientists and materials scientists who consider single interfaces or solids.

The following represents a list of possibilities. Efforts at all scales need to proceed in parallel. In each case it will be important to understand the effect of time and temperature on the results.

Quantum Mechanical Calculations

These calculations include the effects of the electronic structure directly, allowing first-principle studies of chemical changes and electronic excitations. Current techniques can find energies and forces in static, zero-temperature systems with approximately 1,000 atoms. The Carr-Parinello and related schemes allow dynamic, finite-temperature simulations of up to 100 atoms.

Such system sizes only allow ideal, flat, periodic surfaces to be studied. However, several phenomena can only be treated reliably with these techniques. Some of the most important are:

- adhesion and surface energies, calculable from first principles
- interatomic potentials that are included directly in friction models or through quantities to be included in potential fits
- dependence of friction forces on force potentials between atoms inside and between the sliding bodies
- chemistry of single surfaces, formation of oxides, layers, modified structures
- changes in chemistry under sliding; formation and stretching of bonds between the surfaces, transfer of translation energy into heat, scission of lubricant molecules, chemical reactions between surfaces

Molecular Dynamics Simulations

Given interatomic or intermolecular potentials, one may study equilibrium and dynamic behavior of 10^6 to 10^7 atoms and time scales up to a few nanoseconds. These systems are large enough to allow studies of the effect of roughness and to include lubricants. With appropriate potentials it may be possible to model simple chemical changes at these scales. Previous work indicates that continuum mechanics begin to apply at fairly small scales (dimensions of 5 or more atomic diameters). This may allow molecular dynamics to span the gap between atomic and macroscopic processes. Because all particle motions may be followed and analyzed, it is possible to study the generation, propagation, and dissipation of energy in lattice vibrations. Opportunities for study include:

- effect of roughness at the scale of a few atoms
- influence of lubricants or third bodies
- velocity dependence of dry friction
- time dependence of static friction
- generation of dislocations, plastic deformation
- local heating and its consequences
- tribochemistry, formation of tribo-polymers at surface, modification of reaction rates by friction
- microscopic understanding of viscosity and non-Newtonian rheology
- constitutive relations for macroscopic studies

Basic Friction Theories

These explore the conversions of translational energy into phonon excitations and the energy transfers from one phonon mode to a thermal equilibrium bath. These theories presently predict vanishing or extremely small friction forces for many situations, in particular, the case of two incommensurate solids rubbing against each other. The open question at this time is whether such low friction actually exists but has escaped experimental detection (except for case of inert gases physisorbed on crystals), or whether the origin of friction must be sought in different models than presently contemplated. The further development of these theories will identify the essential material properties that determine friction and provide guidelines for the design of materials presenting desired friction properties.

Macroscopic Calculations — Finite Element and Continuum

There is a need to build constitutive relations that go beyond simple coefficients of friction into macroscopic calculations. This would help our understanding of microslips, initiation of sliding, effect of machine compliance and inertia, and other processes discussed in the section on technology. Finite-element calculations of interface contact and interactions still need refinement.

Some opportunities for study by macroscopic techniques include:

- effects of material stiffness and their influence on the degree of adhesion and stick-slip and the control of static friction
- contact deformation and plastic or viscoelastic dissipation of energy (well understood in some cases)
- changes in contact area with time, caused by the flow of lubricant or the viscoelastic response of the surface
- time-dependent loading, chatter, stick-slip
- aggregate friction in granular flow
- solid body motion in contact (slipping and rolling) and collisions

Several important pieces of experimental information would be crucial in determining the type and form of idealized systems that would be studied. They would also provide benchmarks and tests. Items of interest include the surface topography; the variation of the contact and the size and number of contacting asperities with load, time, temperature, and relative movement of the surfaces; and the chemical composition of the surfaces.

Summary of Research Recommendations

- basic theory of friction, relation to material properties supporting science: solid state theory, chemical bonds, interatomic potentials, surface energies
- theory of adhesion
- theory of tribochemical processes, effect of bond stretching and breaking on reactivity (mechanical catalysis)
- molecular dynamics of friction phenomena, including local heating, tribochemistry, modeling of boundary lubricants

OVERVIEW OF THE INTERPLAY BETWEEN THE THEORETICAL CONCEPTS IN FRICTION

Quantum Mechanical

10^2–10^3 atoms
Adhesion
Chemistry
Electronic processes
→ dissipation, emission

Interatomic Potentials

focus concepts

Molecular Dynamics

10^2–10^7 atoms
Roughness
Lubricants
Tribochemistry
Dissipation by phonons

Constitutive Relations

focus concepts

Continuum

Macroscopic
Machines
Mechanical properties
→ stiffness, inertia
Contact deformation
Granular flow

concepts tests

EXPERIMENTS

• • • • • •

ATOMIC AND MOLECULAR VIEWPOINT

Irvin Singer, *Chairman*

In the last 20 years, evidence of the existence of surface films has increased considerably. Instruments for squeezing and shearing of liquids have now measured the properties of some liquid films down to single monolayers. The modeling of molecular structure and atomic interactions with super computers has helped us visualize mechanisms of energy loss in sliding. Further studies of electron charge transfer and emission of energetic particles from straining and fracturing solids have provided evidence that not all energy dissipation occurs directly by phonon creation and that relatively large excitations often accompany friction.

Future descriptions and models for friction will eventually be built on information obtained from several areas of study that generally focus on the physical properties of atoms and molecules. Today these include the following.

• *The squeezing and shearing of known fluids between atomically smooth surfaces, where the interaction of the molecules with the surfaces and the geometric confinement create a unique rheological response of the fluid.* One of the main challenges remaining is to establish a connection with lubrication in applied situations where ultraflat surfaces and very low velocities are not usually present.

• *Friction experiments with microscopic points that model a single asperity (for instance in the atomic force microscope).* They aim at measuring the interatomic forces responsible for friction. These measurements provide the opportunity of probing these forces with variable distance between the surfaces. The techniques have been successful in imaging surfaces, including the behavior of lubricating films (such as those used in computer disks). Fundamental information on friction phenomena, however, is much more difficult to obtain because realistic contact entails wear of the tip and prevents reproducible measurements.

• *Friction experiments using surfaces with controlled geometry and surface composition, aimed at verifying the novel basic friction theories.* This remains almost a totally unmet challenge. Usable contact with the theories requires very simple and well-characterized surfaces; these are the surfaces prepared through the classic techniques of surface science, namely, single crystal surfaces, atomically clean or covered with controlled and known adsorbates in ultrahigh vacuum. There remains an interesting controversy: such surfaces are known experimentally to present very high friction, yet near-zero friction is predicted for all but crystallographically commensurate surfaces.

• *Modeling of the behavior of limited masses of atoms under stress in large computers.*

• *Deformation and fracture studies at very small scales, at which the classic concepts, such as dislocation theory for example, do not apply.*

There must be an effort to translate the findings of atomic studies into terms that are compatible with the macroscopic mechanics approach and to determine where and how these findings are relevant to applications. Those who focus on continuum mechanics should simultaneously adjust their methods to accommodate the specialized language of the fundamental studies. We recommend that new emphasis be placed on effective communication of the results of these studies to continuum mechanics specialists. The latter are the bridge to applications.

In addition, we recommend that further understanding be sought of the rheology of liquids that is relevant to lubrication. The very high shear rates and thin films encountered in many machines are beyond the range of validity of simple viscosity theories, and the potential benefit of viscoelasticity of complex fluids such as liquid crystals or shear-thinning fluids is barely known.

Summary of Research Recommendations

- experiments designed to verify basic theories
- rheology of confined liquids
- deformation and fracture at very small scales (no dislocations)
- better understanding of the rheology of liquids
- bridge between nano-tribology and applications

● ● ● ● ● ●

SURFACE CHEMICAL VIEW OF TRIBOLOGY

Richard S. Polizzotti, *Chairman*

Considering the current state of surface probe development, the key issues are no longer whether we can identify the composition of tribological surface layers, but rather whether we can establish an understanding of the relations between friction and the structure, chemistry, and mechanical properties of surfaces. With that perspective as a background, the key science issues which we have identified as the primary drivers for future research are as follows.

• *Identification of the chemical and physical properties of the surface layers necessary to control friction.* This issue suggests the need for systematic studies of the relation between friction performance and surface characterization, with a particular emphasis on (1) the mechanical properties (shear strength, hardness, toughness, elasticity, adhesion) of the tribochemical film/substrate system; (2) spatially resolved structural and chemical analysis of the reaction product layer; and (3) establishing the kinetics of tribochemical film formation, with an understanding of whether equilibrium or non-equilibrium phases are formed.

• *Development of in-situ/real-time probes of interfaces.* The principal focus of these studies would be the characterizations of the local contact conditions (e.g., temperature, pressure, shear stress), and the in-contact molecular structures and chemical bonding of friction-modifying additives and of tribochemical reaction products.

• *Kinetics and thermodynamics of adsorption, competition for surface sites, and reaction of chemically active lubricant additives.* This information is of fundamental importance in understanding the chemical and structural evolution of the tribochemical layers whose form (as temperature, load, sliding speed, etc.) varies in fluid lubrication.

• *Identification of the physical/chemical properties controlling the magnitude of the interfacial shear strength in dry friction.* In particular, we need to understand whether the bulk mechanical properties of the phases identified in the interfacial region are sufficient to explain the observed behavior, or whether these properties are modified by the presence of the interfaces, and how reactions between the environment (e.g., air, H_2, etc.) and the interfaces affect the interfacial shear strength.

Despite the formidable problems, there is real potential for success. This stems from the recent development of a variety of powerful new tools for in-situ probing of the lubricated interface that is beginning to open new avenues of research. The following are some key tools identified with the greatest potential for impacting the issues described above.

• Advanced light sources, associated with the synchrotron, are making a variety of new spectroscopies possible. In particular, EXAFS/SXAFS (Extended X-ray Absorption Fine Structure/Surface Extended...) allows in-situ element specific determination of local molecular structure and is directly applicable to studies of metal center coordination in metal-organic additives both in solution and as surface adsorbed species. Extension of these techniques to fluorescence emission induced by electron excitation (EXEFLS) in a transmission electron microscope (TEM) can be used to study local sites in amorphous wear debris for particles as small as 100 Å.

• Advances in imaging x-ray photoelectron spectroscopy are making possible the chemical mapping of complex heterogeneous friction/wear contacts with < 500 Å spatial resolution.

• X-ray microscopy employing near-edge electronic fine-structure differences as a contrast mechanism has been shown to be an effective tool for in-situ imaging of interfaces in complex materials.

• In-situ linear and nonlinear optical spectroscopies have been developed and are now being applied to direct measurement of the structure and chemistry of the tribochemical layers found in the contact and the characterization of the fluid structure in the contrived contact as a function of the structure of the fluid molecules. Of particular importance — the development of in-situ FTIR techniques and the development of fluorescent probes, the use of Raman spectroscopies for in-contact measurements of fluid chemistry, pressure, and temperature in the contact; and the development of sum frequency spectroscopy as an interface-specific probe of additive adsorption, structure, and chemical bonding.

Several areas of technical innovation hold promise for relatively short-term impact on the control of friction. In surface modification by energetic beam processing (including ions, electrons, and lasers) and physical and chemical vapor deposition of tribological coatings, the key technological issue limiting broad implementation is cost. The key to continued development is the discovery of niche markets and critical applications where these technologies can provide performance not achievable by other approaches. Vapor and gas delivery of lubricants, taking advantage of tribochemical processes occurring in the contact for in-situ creation of lubricant films, is an innovative approach worthy of further development. Finally, the development of additives for fluid lubrication, in which additive thermal stability and decomposition pathways are tailored to the conditions of the contact for the formation of tribochemical surface films, presents an important approach for study.

Summary of Research Recommendations

- measurement of competitive adsorption of specific additives in the presence of a liquid phase covering a solid (by optical methods)
- further development of the methods for the identification and characterization of thin surface films (composition, structure, mechanical properties)
- adaptation of advanced measurement techniques to tribological experiments

● ● ● ● ● ●

MATERIALS — MICROSCOPIC VIEWPOINT

Stephen Rice, *Chairman*

The microscopic viewpoint of friction as it relates to materials is concerned with phenomena that occur on the scale of asperities. At this scale, gross mechanical parameters such as load and sliding velocity are partitioned into many components, which are accommodated on the scale of local contacts. Each such contact may be considered in terms of load cycling, which ultimately results in changes observable in specimen and counterface materials at the microscopic level. Such changes include the development of surface films or near-surface layers that are chemically distinct from the original materials, as well as the development of deeper zones that are plastically deformed. These triboprocessed near-surface zones are further characterized by the development of voids and cracks, and by the formation of debris particles. The redistribution of asperities and the formation of debris or third bodies due to shear accommodation creates a complex interface that gives rise to changes in the frictional behavior of tribosystems.

The primary questions in this area are as follows. First, what are the specifics of the partitioning of gross mechanical parameters that occurs? For example, what kinds of distributions of contact areas exist at any point in time, and what are the local time-varying stress fields and temperature distributions, and what kind of chemical reactions occur? If the answers to these questions were known, and if a suitable constitutive model were available, then one could begin to address the nature of the resulting triboprocesses (from both the materials and fracture mechanics points of view). Second, how do tribodynamic effects influence the above interactions? It is well known that frictional phenomena are time dependent. For a given laboratory apparatus or actual working machine involving sliding contact, what are the modes of vibration and what are the frequencies and magnitudes of the torsional and longitudinal and transverse displacements? If the displacement magnitudes are significant relative to asperity or debris dimensions, or to the nominal separation between reference planes on the specimen and counterface bodies, then it is clear that contact mechanics cannot be pursued as a quasi-static exercise, especially at the microscopic level. It follows that tribodynamic phenomena must be understood and integrated into analyses in the areas of contact mechanics as well as in studies on chemomechanical triboprocessing of near-surface material layers.

The scope of needed investigations is daunting. However, the list of variables has been increasingly refined (e.g., Czichos). One promising approach is to prioritize such lists in reference to specific engineering applications. For any such application, there are system constraints, leaving a subset of parameters to be studied to achieve optimal results. In some systems, it may be possible to separate adhesive and plowing components of friction, and to minimize both, perhaps by intentional use of third bodies. Indeed, the role of third bodies in affecting each of these components could be usefully studied. Likewise, it is possible that new experiments using single-crystal materials could be designed to elucidate some basic features of frictional phenomena; this area may be relevant to emerging microelectromechanical systems. In addition, for some materials under given sets of applied loads and speeds, there is the possibility that rheological properties of very thin films (primarily studied between ultraflat surfaces at low velocity) could be exploited for practical systems. When combined with surface treatments and/or modification techniques (e.g., ion implantation, laser glazing) or surface coatings (e.g., chrome plate), such films should yield significant reductions in friction. When further combined with studies of the role of chemomechanical factors, additional progress is possible. Still more insight will follow when techniques are developed to determine material properties under nonequilibrium conditions of tribocontact (e.g., actual temperatures, stresses, strains, and strain rates). Is it possible to study the rheology of asperities, debris, and near-surface zones in situ? In addition, new techniques are needed to determine time-varying contact areas and measure adhesion. Perhaps new tools, such as the scanning acoustic microscope, the confocal microscope (scanning optical profilometer), the atomic force microscope, and metrology, such as laser speckle, can be applied in such investigations. It is necessary to introduce some of the realities of non-uniform materials making contact in dynamic chemomechanical environments into the computer modeling of tribological behaviors.

Some tribodynamic studies may yield immediate practical results. For example, it may be feasible to exploit vibrations in machining operations to achieve improved surface finish and/or higher efficiencies. Similarly, tribodynamic modeling may allow the prediction of maximum and minimum values of friction coefficient. Such information could be important in design. A further area of interest is the use of dynamic friction data to infer attributes of the interface (e.g., distribution of third bodies). Such information could be exploited in machine monitoring. A systematic study is needed on how various surface treatments and coatings affect friction. Possibilities include diamond coatings on SiC and zirconia on Ti, and the use of Buckminsterfullerine structures in tribological applications. Such studies could lead to design information where coefficients of friction could be predicted or selected.

In conclusion, the following areas are particularly promising:• studies to elucidate actual contact conditions (local area dimensions) and the a sociated material response (triboprocessing of materials under given conditions of local temperatures, stresses, strains, and strain rates)

- projects to investigate novel design approaches to achieve tribological outcomes through surface or interface modification
- a teamed study in which molecular modelers theoretically predict material formulations that appear to offer low adhesion; material scientists then synthesize the materials and test them for frictional performance
- study to determine actual material properties of asperities inside a tribocontact (solid rheology)
- a study to determine dynamic behavior of actual tribosystem comprised of two constrained sides (specimen and counterface) in which time-varying frictional and normal forces are predicted, and which allows investigation/study of the role of dynamic fluctuations (normal and transverse displacements) or asperity and debris interaction in the interface

● ● ● ● ● ●

MATERIALS — CONTINUUM VIEWPOINT

Andres Soom, *Chairman*

The description of tribological contacts from the continuum mechanics point of view extends from the microscopic (asperity level or smaller) to more global descriptions of the entire interface region, including the macroscopic behavior of contact forces between sliding or rolling bodies.

At the contact level, be it concentrated or extended, there is a need to model the entire contact region, which includes the real interfaces, the deformed surface texture, and some near-surface layers. The thickness of this region is generally of the order of the composite surface roughness and includes all material whose mechanical behavior is different from the bulk in terms of strength, effective elastic modulus, mechanical dissipation, and thermal properties.

Of particular importance is the connection of the parameters of such models to experimental data that could be further correlated with design and manufacturing variables. There is a need to view the tribological interface region as more than just a frictional boundary condition. The further development of such approaches, some of which exist in various forms in the literature, may help bring more realistic contact models into the finite element computations and other analyses frequently performed in connection with mechanical design.

In the above context, the modeling of mixed lubrication and the changes in surface texture, material properties, and chemical composition that occur during running in are important areas in which our current knowledge is lacking.

Another topic of practical importance where more work is needed is static friction, along with transitions between sticking and slipping. In this context, microslip and other localized slip, in the presence of surface films and liquid lubricants, is of interest. Contact motions and multiaxial system dynamics must usually be included in measurement and modeling.

Many contacts operate under unsteady or dynamic conditions. Macroscopic models of friction, including parametric dependencies not only on sliding speed, but load, surface texture, and lubricant properties when present are needed for motion control, machine dynamic analysis, and noise and vibration applications.

The simultaneous application of Euclidean and fractal geometry or the possibility of including localized events, e.g., hot spots, into more macroscopic models could be of value.

Tribological principles should be assembled and made accessible so that they can be applied more broadly to engineering systems and improve performance and quality in many areas (e.g., sports, consumer products, space, manufacturing) where our existing knowledge is not being systematically applied.

Educational material needs to be developed for design engineers. In order to be used, this information must be concise, address specific design questions, and be expressed in the language of the user.

Summary of Research Recommendations

- dynamics of asperity level interactions in dry and lubricated contacts
- continuum mechanics modeling of interface regions, coupled with experimental studies
- static friction and stick-slip transitions
- unsteady friction models for machine dynamics, motion control, and vibration problems
- novel geometrical descriptions of surfaces and contacts
- tribological design principles
- development of effective educational material that can be used to transmit these principles to design engineers

ORGANIZATION AND EDUCATION

With the development of new techniques and concepts in very different scientific disciplines, theoretical and experimental physics and chemistry, and mechanical engineering and materials science, the interdisciplinary nature of friction research presents a challenge. It is imperative to maintain contact and collaboration between the efforts in nanotribology and those in the more applied areas of the field. This makes stringent demands on the interdisciplinary skills — it requires that the researcher be reasonably conversant with the work in the other disciplines, if not necessarily with the details of the work, at least with the basic concepts of the method and the significance of the results.

A second interdisciplinary challenge is presented by the utilization of tribology in the design, manufacture, and operation of machines and systems. In this case, the precise, systematic, and concise presentation of the concepts and facts must be made in a technical language close to that of the designer.

Review of the Recommendations for Support by the National Science Foundation

We review here the recommendations made with each of the topics covered in detail.

A Research Recommendations Based on Technological Needs

A1 Extension of the temperature range in which friction is controlled, from cryogenic to very high temperatures. There is a need for moderate increases, possibly still in the range of advanced liquid lubricants and extreme temperatures, where novel schemes must be found.
Served by A4, B4, B5, B8, B9

A2 Friction control (lubrication) in difficult chemical environments (human body, alternate fuels, processed fluids)
Served by A4, B4, B5, B8

A3 Environment friendly lubricants in automotive and metal-forming industries
Served by A4, B4, B5

A4 Novel chemisorbed lubricant molecules (boundary lubricants)
Served by B4, B5, B8

A5 Friction control (lubrication) adapted to novel materials (ceramics, polymers, fibers)
Served by B4, B5, B8, B9

A6 Novel methods for control of friction of surfaces in very close contact

A7 Better understanding of static vs. dynamic friction (e.g., in rolling)
Served by B1, B2, B6, B7, B11, B12

A8 Role of surface topography in lubricated systems, especially in mixed lubrication
Lubrication at low speed, sheet forming, control of run-in of machines, control (avoidance) of stick-slip and static friction
Requires further research in microelastohydrodynamic lubrication

A9 Friction control of unlubricated surfaces
(For instance the increased traction in railroads or the development of low-friction aterials)
Served by B1, B2, B10, B11

A10 Better mathematical models of friction for use in the design of machines and processing
Served by B1, B2, theories of lubrication

A11 Relationship between friction and machine dynamics
Control of system vibrations
Served by B1, B2, B7, B11, B12

A12 Information transfer in tribology
Tribological criteria for machine design
Educational material for engineers

B Research Recommendations in the Basic Science of Friction

B1 Theory
Fundamental energy dissipation processes
Interatomic force potentials
Computer experiments (molecular dynamics)

B2 Controlled experiments supporting theoretical developments
Relation of material properties to friction

B3 Atomic and molecular friction and lubrication
Dynamics of confined molecules
Thin-layer rheology and its relation to viscosity
Shear-induced chemistry

B4 Surface chemistry
Adaptation of new surface analytical tools to tribology
Tribochemical reactions of sliding bodies with the environment and lubricants
Chemisorption of antifriction or antiwear additives on surfaces

B5 Tribochemistry
Friction-enhanced kinetics of reactions

B6 Third body effects on friction

B7 Control of vibrations
Their effect on friction and their dependence on velocity

B8 Prediction and generation of tribological films
Coatings
Self-generation

B9 Design and application of new tribomaterials
Control of friction

B10 Engineered interfaces
Alloys that form low-friction films
Implantation and coating for low friction
Texturing of surfaces

B11 Relationship between macroscopic and microscopic viewpoints

B12 Surface dynamics
Behavior of asperities in sliding
Use of vibrations for friction control

Suggested Research Programs

The technological needs and the scientific advances and opportunities outlined in the preceding section form a number of coherent research programs warranting consideration for support by the National Science Foundation. These are outlined below.

These possible programs evolved from the workshop and represent a picture of the present situation. Quite probably, there are some omissions, which some will find essential. The combinations of problems are, of course, somewhat arbitrary and omit an essential element of research – namely the truly novel, the unexpected idea or discovery. It is hoped that in a few years such innovations will make the preceding set of recommendations obsolete.

One of the important developments is the widening of the interdisciplinary scope of research in friction, in particular through the introduction of theories of friction based on solid state physics, the molecular dynamics simulations, and the development of optical techniques for the detection of surface chemistry. It is considered imperative for progress that the different activities be integrated in an interdisciplinary program. This presents a specific technical challenge: such integration and collaboration between practitioners of different disciplines requires the mutual understanding of concepts and results.

Such an understanding requires a degree of familiarity with different languages and techniques; the difficulty of acquiring the required broad-based knowledge should not be underestimated and creative efforts in this direction should be encouraged.

I Fundamental Science of Friction

Development of a theoretical understanding of the mechanisms by which kinetic translation energy is transformed into other, ultimately thermal, energy; identification of the relevant material properties determining friction

Experiments designed specifically to verify the theories

Molecular dynamic modeling of the same phenomena

Application of the knowledge acquired:

- to design low-friction materials or materials with controlled friction
- for a better control of static vs. dynamic friction, vibration phenomena
- to establish mathematical models of friction
- to explore novel friction-control schemes in high density
- for the design and application of new tribomaterials

II Chemistry of Lubricants

Design and synthesis of novel lubricant fluids to extend the useful temperature range. (This may be an important but elusive goal.) Design and synthesis of novel boundary lubricants.

Application to:
- high-temperature lubrication
- high-temperature lubrication by gas phase transport of lubricant
- lubrication for magnetic storage
- pollution-free lubricants

III **Surface Chemistry of Lubrication**

Science of adhesion

Development of novel methods for the study of adsorbed molecules in the presence of a liquid

Tribochemistry

Application to:
- high-temperature lubrication
- friction control in difficult environments
- friction control adapted to novel materials
- friction control in data processing and micromechanics
- prediction and control of tribological films
- lubrication of biomaterials and prostheses

IV **Rheology of Lubricants and Materials**

Science of fluid rheology (temperature dependence, viscoelasticity, behavior in high shear rates)

Rheology of fluids confined in very thin layers

Rheology of fluids in the vicinity of surfaces

Rheology of powders

Application to:
- high-temperature lubrication
- lubrication by novel, complex fluids
- powder metallurgy and processing of ceramics and composites

V **Advances in Micro-Elasto-Hydrodynamic Lubrication**

Lubrication and friction as a function of surface topography

Application to materials processing, in particular, rolling

VI **Mechanics**

Dynamics of friction (time-dependence, vibration generation, stick-slip)

Dynamics of mechanical systems with friction vibration

Application to:
- machine chatter
- friction reduction by vibration
- computer modeling of friction

VII **Material Science of Friction and Wear**

Behavior of asperities in friction

Wear and material deformation in friction

Fatigue
Application to:
- running in
- rolling friction

VIII Study of Biological Friction Phenomena
Application to lubrication of implants, prostheses
Development of novel lubrication schemes (biomimetic lubrication)

Workshop Program

Wednesday, June 24, 1992

Morning:	Registration
12:00	Lunch
1:20	Purpose of the Workshop (T. E. Fischer and S. Jahanmir)
1:30	Overview of Problems and Concepts in Friction (D. Tabor)
2:30	Overview of the NATO Advanced Summer Institute (I. L. Singer)
3:30	Break
4:00	Overview of Industrial Needs (M. Peterson)
5:00	General Discussion
5:30	Informal Discussions (Cash Bar)
6:00	Dinner
7:00	Keynote Address: "An Overview of Technology Policy in the Department of Commerce," Deborah Wince-Smith, Assistant Secretary of Technology Policy (tentative)

Thursday, June 25

8:00	Discussion Groups — Technological Needs 1. Mechanical Systems — Engines, Turbines, Transportation, Machinery (Chairman: Robert Harmon) 2. Materials Processing and Manufacturing (Chairman: Harmon Nine) 3. Data Processing and Micromechanical Systems (Chairman: Larry Spector) 4. Road and Rail Friction Chairman (Tom Yeage) 5. Other Applications and Emerging Technologies (Chairman: Frederick Ling)
10:30	Break (Chairmen Prepare Presentations)
11:00	Presentation of the Findings of the Groups and General Discussion
12:00	Lunch
1:30	Discussion Groups — Scientifc Needs and Opportunuties 1. Fundamental Theories (Chairman: Marc Robbins) 2. Atomic and Molecular Scale (Chairman: Garry McClelland) 3. Materials — Microscopic Viewpoint (Chairman: Steve Rice) 4. Materials — Continuum Viewpoint (Chairman: Andres Soom) 5. Surface Chemical Viewpoint (Chairman: Richard Polizzotti)
4:00	Break (Chairmen Prepare Presentations)
4:30	Presentations of the Findings and General Discussion
5:30	Informal Discussions (Cash Bar)
6:30	Dinner
8:00	General Discussion
9:00	Adjourn

Friday, June 26

8:00	Discussion Groups — Formulation of Recommendations The aim of all groups is to combine the findings of basic science opportunities and technological needs identified the previous day and formulate recommended research programs. (Moderators: T. E. Fischer, S. Jahanmir, M. Peterson)
10:00	Break
10:30	Report of the Chairmen and Discussion
12:00	Lunch
1:00	Adjourn, Discussion Chairs Write Report
4:00	Discussion Chairs Adjourn

List of Attendees

Mr. Serguei Altynov
Ohio State University
116 West 19th Street
Columbus, OH 43210

Mr. Richard Colton
Naval Research Laboratory
Washington, DC 20375

Dr. Howard E. Clark
ASME, Suite 906
1828 L Street NW
Washington, DC 20036-5402

Mr. Marshall L. Devenpeck
Alcoa
FABT–ATC–B
Alcoa Labs
Alcoa Center, PA 15069–0001

Dr. Traugott E. Fischer
Professor
Department of Materials Science
and Engineering
Stevens Institute of Technology
Castle Point Station
Hoboken, NJ 07030

Mr. Suresh Goyal
AT&T Bell Laboratories
Room 2D 404
600 Mountain Avenue
Murray Hill, NJ 07974

Mr. Peter J. Blau
Metals and Ceramics Division
Oak Ridge National Lab
P.O. Box 2008
Oak Ridge, TN 37831–6063

Dr. Daniel B. Dawson
Division 8746
Sandia National Labs
P.O. Box 969
Livermore, CA 94556

Dr. Sarah Dillich
U.S. Bureau of Mines
2401 E. Street NW
Washington, DC 20241

Mr. Pawan Goenka
Fluid Mechanics Department
General Motors Research Labs
30500 Mound Road
Warren, MI 48090–9055

Mr. Robert Harmon
Consultant
25 Schalren Drive
Latham, NY 12110–1229

Mr. George Hosang
Sunstrand Power Systems
4400 Ruffin Road
P.O. Box 85757
San Diego, CA 92186–5757

Dr. Said Jahanmir
NIST
Building 220, Room A215
Gaithersburg, MD 20890

Dr. Jorn Larsen–Basse
Program Manager, Tribology
National Science Foundation
Room 1108
1800 G Street NW
Washington, DC 20550

Mr. Jeffrey R. Lince
Aerospace Corporation
P.O. Box 92957
Los Angeles, CA 90009–2957

Professor Jacob Israelachvili
Dept. of Chemical & Nuclear Engineering
University of California
Santa Barbara, CA 93106

Dr. Elmer E. Klaus
Department of Chemical Engineering
Penn State University
108 Fenski Laboratory
University Park, PA 16802–4400

Professor James Lauer
Department of Mechanical Engineering
Rennselear Polytechnic Institute
Troy, NY 12181

Dr. Frederick F. Ling, P.E.
President
Institute for Production Research
Room 8–E
488 Seventh Avenue
New York, NY 10018

Mr. Harmon Nine
Fluid Mechanics Department
General Motors Research Labs
30500 Mound Road
Warren, MI 48090–9055

Dr. Joseph Perez
Office of Transportation Materials
Department of Energy
1000 Independence SW
Washington, DC 20585

Dr. Richard S. Polizzotti
Exxon Research and Engineering Co.
Annandale, NJ 08801

Professor Marc Robbins
Physics Department
Johns Hopkins University
3400 North Charles Street
Baltimore, MD 21218

Dr. Irwin Singer
Naval Research Laboratory
Code 6176
Washington, DC 20375

Dr. Jacob Sokoloff
Physics Department
North Eastern University
Boston, MA 02115

Mr. Marshall B. Peterson
Wear Sciences, Inc.
925 Mallard Circle
Arnold, MD 21012

Mr. Steven Rice
College of Engineering
University of Central Florida
P.O. Box 25000
Orlando, FL 32816–2993

Mr. Lewis B. Sibley
Tribology Consultants, Inc.
504 Foxwood Lane
Paoli, PA 19301

Dr. David P. Smith
Division Scientist
Data Storage Products Division
3M Company
Building 236–GD–18
St. Paul, MN 55144–1000

Professor Andres Soom
Mechanical & Aerospace Engineering
State University of New York
Buffalo, NY 14214

Dr. C. J. Spector, Manager
Recording Physics
IBM
5600 Cottle Road, E19, Building 28
San Jose, CA 95193

Mr. Steve Thomas Quaker
Chemists
Elm and Lee Streets
Conshohoken, PA 19428

Mr. David Tabor
Cavendish Laboratory
Madingly Road
Cambridge 6CB3OME
United Kingdom

Dr. James J. Wert, P.E.
Professor of Metallurgy
Vanderbilt University
Box 1621, Station B
Nashville, TN 37225

LIST OF PUBLICATIONS
ASME CENTER FOR RESEARCH AND TECHNOLOGY DEVELOPMENT

Volume No.	Title
CRTD-1	*Consensus on Current Practices for Lay Up of Idustrial and UtilityBoilers,*1985, 28 pp. Book Number H00336
CRTD-2	*Research Needs in Thermal Systems,* 1986, 221 pp. Book Number I00212
CRTD-3	*Goals and Priorities for Research in Engineering Design,* 1986, 164 pp. Book Number I00230
CRTD-4	*Pressure Systems Energy Release Protection (Gas Pressurized Systems),* 1986, 425 pp. NASA Contractor Report 178090
CRTD-5	*Thermodynamic Data for Biomass Materials and Waste Components,* 1987, 496 pp. Book Number H00409
CRTD-6	*Analysis of the Feasibility of Replacing Asbestos in Automobile and Truck Brakes,* 1988, 136 pp. Book Number I00264
CRTD-7	*Hazardous Waste Incineration: A Resource Document,* 1988, 192 pp. Book Number I00266
CRTD-8	*Research Needs in Dynamic Systems and Control – Volume 1: Strategic Initiatives and Opportunities,* 1988, 58 pp. Book Number I00275
CRTD-9	*Research Needs in Dynamic Systems and Control – Volume 2: Acoustics and Noise Control,* 1988, 48 pp. Book Number I00276
CRTD-10	*Research Needs in Dynamic Systems and Control – Volume 3: Control of Mechanical Systems,* 1988, 40 pp. Book Number I00277
CRTD-11	*Research Needs in Dynamic Systems and Control – Volume 4: Machine Dynamics,* 1988, 37 pp. Book Number I00278
CRTD-12	*Research Needs in Dynamic Systems and Control – Volume 5: Nonlinear Dynamics,* 1988, 32 pp. Book Number I00279
CRTD-13	*Peer Review Thermal Energy Storage Program,* 1989, 46 pp. Available from CRTD, Washington, D.C.
CRTD-14	*The ASME Handbook on Water Technology for Thermal Power Systems,* 1989, 1900 pp. Book Number I00284
CRTD-15	*Research Needs and Technological Opportunities in Mechanical Tolerancing,* 1990, 51 pp. Book Number I00298
CRTD-15-1	*Selected Case Studies in the Use of Tolerance and Deviation Information During Design of Representative Industrial Products,* 1992, 156 pp. Book Number I00329
CRTD-15-2	*Mathematization of Dimensioning and Tolerancing,* 1992, 38 pp Book Number I00337
CRTD-16	*Engineering Fluid Mechanics Workshop Report,* 1990, 114 pp. Book Number I00299

LIST OF PUBLICATIONS
ASME CENTER FOR RESEARCH AND TECHNOLOGY DEVELOPMENT

Volume No.	Title
CRTD-17	*ASME Research Committee on Corrosion and Deposits From Combustion Gases Seminar of Fireside Fouling Problems,* 1990, 183 pp.
CRTD-18	*ASME Ash Fusion Research Project,* 1990, 11 pp. Book Number I00307
CRTD-19	*Achievements in Tribology,* 1990, 173 pp. Book Number H00269
CRTD-20-1	*Risk-Based Inspection – Development of Guidelines: Volume 1, General Document,* 1991, 155 pp. Book Number I00311
CRTD-20-2	*Risk-Based Inspection – Development of Guidelines: Volume 2, Part 1, Light Water Reactor (LWR) Nuclear Power Plant Components,* 1992, 156 pp. Book Number I00321
CRTD-21	*Hazardous Waste Destruction by High Temperature Incineration,* 1991, 21 pp. Book Number I0266A
CRTD-22	*Mining Workshop for Nuclear Waste Cleanup,* 1991, 145 pp. Book Number I00314
CRTD-23	*The Use of Decision-Analytic Reliability Methods in ASME Codes and Standards Work,* 1993, 120 pp. Book Number I00351
CRTD-24	*ASME/Bureau of Mines Investigative Program on Vitrification of Residue From Municipal Waste Combustion Systems, (to be published)*
CRTD-25	*Installation of Plastic Gas Pipeline in Steel Conduits Across Bridges,* 1993, 64 pp. Book Number I00353
CRTD-26	*Lubrication Technology for Advanced Engines – An Assessment of Industrial Needs,* 1993, 104 pp. Book Number I00352
CRTD-27	*1993 International Forum on Dimensional Tolerancing and Metrology,* 1993, 324,pp. Book Number I00360
CRTD-28	*Research Needs and Opportunities in Friction,* 1993, 48, pp Book Number I00365